YOUR KNOWLEDGE HAS VALUE

- We will publish your bachelor's and
  master's thesis, essays and papers

- Your own eBook and book -
  sold worldwide in all relevant shops

- Earn money with each sale

Upload your text at www.GRIN.com
and publish for free

Bibliographic information published by the German National Library:

The German National Library lists this publication in the National Bibliography; detailed bibliographic data are available on the Internet at http://dnb.dnb.de .

Imprint:

Copyright © 2017 GRIN Verlag, Open Publishing GmbH
Print and binding: Books on Demand GmbH, Norderstedt Germany
ISBN: 9783668511828

This book at GRIN:

http://www.grin.com/en/e-book/374590/safety-of-aspartame

Aslı Ucar, Mustafa Özgür, Serkan Yilmaz

Safety of Aspartame

GRIN Publishing

SAFETY OF ASPARTAME

Aslı UÇAR[1], Mustafa ÖZGÜR[1], Serkan YILMAZ[2]

[1]Ankara University, Faculty of Health Sciences, Department of Nutrition and Dietetics, Ankara, Turkey

[2]Ankara University, Faculty of Health Sciences, Department of Midwifery, Ankara, Turkey

1. INTRODUCTION

Sweeteners are the most discussed among the food additives. Those, used as alternatives to sucrose, are generally referred to as "alternative sweeteners" (Mortensen, 2006). The first registered sweetener was honey (Bright 1999; Weihrauch and Diehl 2004), but as time pasts the common sugar took its place. Artificial sweeteners came into use, because diabetes and obesity rate was increased due to use of common sugar. The first used artificial sweetener was saccharin (Bright 1999; Weihrauch and Diehl 2004). Aspartame and cyclamate were used following the saccharin.

They are produced to be used instead of sugar, have the same taste as sugar, are not considered as harmful to health, have low calorie and/or without calories (Position of the American Dietetic Association, 2004). Increasing with the prevalence of obesity, individuals wishing to reduce energy intake have talented particularly to energy-free sweeteners and low-calorie products (World Health Organization, 2008). The products made with sweeteners are equivalent to the product made with sugar being preferred by producers and consumers (Blackburn and et al., 1997). During the past two decades, worldwide low-calorie food consumption has considerably increased, thus leading to an increase in health concerns associated with high intake of synthetic sweeteners (Bergamo et al. 2011).

"Intense sweeteners" are the sweeteners that produce the required effect in minute quantities, because of their intense sweetness. To emphasize that most of them are produced by chemical synthesis, some call them "artificial" sweeteners, whereas sucrose and other sugars naturally found in plants are considered "natural" (SCF, 1985). Also depending on being source of calories, they have been classified as nutritive and non-nutritive (Whitehouse et al. 2008). While xylitol, sorbitol, mannitol and erythritol are energy-containing artificial sweeteners, acesulfame potassium (asesulfame-K), aspartame, saccharin, cyclamate and sucralose are non-energy artificial sweeteners. The sugar alcohols xylitol, sorbitol and mannitol give an average energy of 2.4 kcal/1gram (Tüfekçi, 2014).

Worldwide, in the 90s it was found that the use of saccharin, cyclamate and aspartame accounted for 7.19%, it constituted 9.05% in 2000 and 9.60% in 2010 (MECAS, 2012). Of the artificial sweeteners, aspartame represents 62% of the sweetener market in terms of global consumption (Fry, 1999) In Turkey, the share of saccharin, cyclamate and aspartame in the entire sweetener market was 6% in 2006 and about 11% in 2013 (TC Sugar Corporation, 2015). The general characteristics of the non-energy artificial sweeteners and the amounts allowed for daily use (ADI) found in Turkey are given in Table 1 (Tufekci, 2014).

Table 1. General characteristics of artificial sweeteners and amounts allowed for daily use

Sweeteners Type	Sweetness (sucrose =1)	Sweet Character	Maximum Dose (mg / kg / day)
Acesulfame-K	130-200	Similar the sucrose, is perceived quickly, does not leave a different taste in the mouth. High quantities produce a sense of bitterness. Resistant to the heat	15
Aspartame	180-200	Similar the sucrose, slightly late sensed, does not leave bitter taste in the mouth. Unstable to the heat	40
Saccharin	200-700	It is detected late. It leaves a lasting taste in the mouth. High quantities produce a sense of bitterness. Resistant to the heat	2.5
Cyclamate	30	Similar the sucrose, leaving a bitter and metallic taste in the mouth. Resistant to the heat	11

| Sukraloz | 600 | The chemical structure is modified to obtain sucrose. It is not absorbed and is thrown away by gaita. Resistant to the heat | 5-15 |

The persistence of artificial sweeteners on the agenda and the desire to work on new sweeteners is due to the desire to achieve a near-sweet taste. There are many popular foods available because artificial sweeteners are assumed to reduce sugar consumption and caloric intake considerably (Humphries et al., 2007) Although non-nutritive or artificial sweeteners are using instead of sucrose in foods, in order to increase flavor and reduce calories, their safety has been controversial (Whitehouse et al., 2008). More people of all ages are choosing to use these products, because the public health attention has turned to reversing the obesity epidemic in the world. These choices may be helpful, for the people who cannot tolerate sugar in their diets (e.g., diabetics). It is also becoming increasingly important as part of the nutritional guidelines for diabetes because of the increased incidence in developed countries as well as in developing countries (Humphries et al., 2007). However, even though sugar alcohols are not a problem for diabetic patients, excessive amounts may have a laxative effect. In the United States, the Food and Drug Administration (FDA) has approved to use four of the non-energy artificial sweeteners (saccharin, aspartame, acesulfame-K and sucralose) (FDA, 2015).

However, there are lots of contradictory studies about relation between sweeteners and diseases (Whitehouse et al., 2008). The companies that produce artificial sweeteners, have conducted many studies about these sweeteners and these are not generally available to consumers. However, artificial sweeteners have associated with health disorders such as hepatotoxicity (Negro et al., 1994), and cancers (Whitehouse et al., 2008), by the result of some independent research studies. However, there is still a vast debate concerning about this issue. The role of sweeteners in the risk of cancer, has been extensively controversied (Bosetti et al., 2009).

In general, sweeteners are substances having a sweet taste. In the past, the Scientific Committee on Food was the scientific guarantor for the safety of food additives (including sweeteners) which are using within the European Union (EU). This responsibility belongs to the European Food Safety Authority, at present. The safety of all sweeteners which are allowed for using in food in the EU, has demonstrated in extensive scientific research. Their safety is documented by the results of tests in humans, several *in vitro* and *in vivo* animal studies, and in some cases epidemiological studies (Mortensen, 2006). Although the scientific evidence shows that the sweeteners which are allowed for using in food are safe, some individuals and organizations remain sceptical about long-term health risks because of their consumption (Mortensen, 2006). Undoubtedly, the most controversial artificial sweetener is aspartame, because of its potential toxicity (Whitehouse et al., 2008), and carcinogenicity (Soffritti et al., 2006), even at a dosage level approximating the ADI for humans (Whitehouse et al., 2008). In this study, you can find aspartame safety in every direction.

2. ASPARTAME

2.1 History

The body mass index has increased from 28.8% to 36.9% in men and from 29.8% to 38.0% in women between 1980 and 2013 in worldwide. It was reported that an action plan should be prepared by the World Health Organization (WHO) to stop the rising prevalence of obesity (Vandevijvere et al., 2015). People are beginning to demand energy-reduced products to reduce energy intakes. With this demand, the food industry has produced oil-reduced milk and dairy products, oil-reduced crackers, natural and artificial sweeteners without sugar. Two scientists from Johns Hopkins University discovered saccharine that was 300 times sweeter than non-nutritious sugar in 1879. Saccharin was used as a sweetener in canned foods until 1907 but later it was banned by the Food and Drug Administration (FDA) (Kalkhoff and Levin, 1978).

Cyclamate has been produced due to safety concerns of saccharin. Cyclamate was isolated in the laboratory of Abbott in 1937 by a student working on antipyretic drugs and was allowed by the FDA to use it in foods in 1951 (Bopp et al., 1986). Cyclamate is used in products such as canned foods, pastries, bacon, toothpaste, gargle products and cereals and consumption of diet sodas containing cyclamate has increased rapidly and has doubled in 1967. However, after studies showing bladder tumors in laboratory rats in high doses of cyclamate then it was banned in 1969 (Administration, 1980). For this reason, international pharmaceutical companies have begun to work to discover new sweeteners.

In 1965, a chemist at G. D. Searle who was studying new treatments for gastric ulcers used a tetrapeptide, which is normally produced in the stomach, to test new antiulcer drugs. While the chemist was making an intermediate, aspartyl-phenylalanine methyl ester, which is one of the steps of synthesizing tetrapeptide, a small amount of the compound accidentally landed on the his hand. Later, without noticing the compound on his hand, the chemist licked his finger and noticed a sweet taste. He realized the material was from the powder aspartyl-phenylalanine methyl ester and it would be safe to taste the material. So he tasted the intermediate again and discovered the sweetness was indeed from aspartame (Mazur, 1984). When aspartame was crystallized from ethanol, it was suggested that the sweet taste of the mixture came from aspartame (Furia, 1972).

It was reported for the first time in 1969 in the Journal of the American Chemical Society that aspartame is an artificial sweetener having 180-200 times more sweet tastes depending on the concentration of sugar (Mazur et al., 1969).

Searle performed two separate toxicity studies on rats on 78th week in 1970 and later examined the pathology and autopsy results of rats in 1972. When the results of the studies are examined, it has been determined that aspartame causes tumor development in the liver, testes and thyroids of rats. When the results of the study were analyzed in another laboratory, it was emphasized that the increase in thyroid and testicular tumors was statistically significant but the increase in liver tumors was not significant and it was approved by the FDA in 1973 (Sarett, 1973). At the end of the 1960s and early 1970s, the prohibition of saccharin and cyclamate led to the need for a new sweetener. Searle started to use aspartame as a flavoring agent in certain foods in 1973 when it received the patent of aspartame from the FDA (Randolph, 1973).

It was determined that aspartame was caused by genetic damage in 13 genetic studies after 15 months of FDA approval, but Alexander Schmidt, who is on the FDA committee, stated that it is appropriate to be used only in dry foods at an panel and provide strong support the current name of the Center for Food Safety and Applied Nutrition (CFSAN). However, due to the objections made, the approval was stopped and it was decided that further studies should be done.

In 1979, the Public Board of Inquiry of FDA set up a study to investigate the safety of aspartame decided that it should not be approved before further research conducted on animal brain tumors. (Smyth, 1983). Despite the contradictory conclusions of the studies and the different opinions of the authorities, the use of aspartame in dry food was reauthorized in 1981 (Public Health Service Food and Drug Administration, 1981)

In July 1983, the National Soft Drink Association (NSDA) stated that aspartame is very volatile in liquid form, and therefore, more works should be done before confirmation of aspartame use in gaseous beverages, but aspartame containing gaseous beverages were launched for consumers (Spingola, 2015).

In 1996, aspartame was approved to be used as a general-purpose artificial sweetener in any kind of food and drink (FDA, 2006).

Since then, aspartame has been widely used in more than 6,000 products, roughly 500 pharmaceutical products, including children's medicines, and has been consumed by hundreds of millions of people in countries all around the world (Butchko & Stargel, 2001).

2.2 Chemical Properties

Aspartame (L-aspartic acid, L-phenylalanine methyl ester) is an artificial sweetener non-nourishing dipeptide which is not found naturally in foods and beverages. There are two forms of aspartame, an alpha and a beta form. The sweet one is only alpha, and when it doesn't specified, "aspartame" here will always refer to the alpha form (Magnuson et al., 2007). It is classified as a non-nutritive sweetener because of its intense sweetness and its ability to be used in very small quantities to sweeten the foods (FDA, 2007). Aspartame, by weight, provides the same calorie intake as sugar (i.e., 4 kcal/g), it can be added at almost 200 times lower levels and get the same sweetness, so providing a far lower net caloric intake. This feature has resulted in the use of aspartame as a low calorie or non-nutritive sweetener in foods and drinks worldwide (Magnuson et al., 2007).

Aspartame is not soluble in fats or oils, but it is slightly soluble in water and ethanol (Magnuson et al. 2007). It is not suitable for cooking or baking, because it is unstable at long-term high heat. But in dry form, it is very stable. It is also unstable in aqueous solutions which slowly return to diketopiperazine (DKP). This causes the sweet taste to be lost. The flavor of Aspartame is like sugar and it improves some tastes. The combinations of it with other intense sweeteners, e.g. saccharin and/or cyclamate, taste sweeter than expected from the sum of the individual sweeteners (Mortensen, 2006).Aspartame is used in cold beverage mixes, chewable multi-vitamins, breakfast cereals, chewing gum, puddings, carbonated drinks, iced drinks, yoghurt and in pharmacy. The European Food Safety Authority (EFSA) assessed the safety of aspartame and included in food additives with E951 code (SCF, 2002).

A large part of the studies related to aspartame was carried out to assess whether it is safe for humans. It is concerned that aspartame-forming compounds will pose a risk in phenylketonuria heterozygotes, newborns, pregnant and lactating mothers, and individuals with Chinese Restaurant Syndrome (Stegink, 1987).

The aspartame metabolite phenylalanine is an amino acid found naturally in the breast milk of mammals; however to those who born with phenylketonuria (PKU), a metabolic disorder caused by an hereditary mutation in the phenylalanine hydroxylase (PAH) gene that prevents the correct metabolization of phenylalanine, high levels of phenylalanine are a health danger. This results in a deleterious accumulation of the amino acid, which causes developmental defects, mental retardation and seizures (Blau et al, 2010). Normal mammalian plasma levels of phenylalanine are approx. 30–50 mM (0.5–0.8 mg/dL), however, 1 out of 50 individuals for the mutation in the phenylalanine hydroxylase gene, is heterozygous (Scriver et al, 1996), the fasting plasma phenylalanine levels were increased compared to the non-carriers (Griffin et al, 1973), with a decrease in the cholinergic ratio of phenylalanine after intravenous loading (Jagenburg et al, 1977). Repeated doses of 8 servings of aspartame-sweetened beverages by PAH heterozygous individuals led to plasma phenylalanine levels of up to 165 mM (Stegink et al, 1990), but remained below levels reported to cause neurotoxicity during acute administration in primates (Collison et al, 2012).

High doses of aspartame and its metabolites have been tested in humans and other animals. It has been proven that not only the metabolites of methanol but also methanol itself are toxic to the brain (Jeganathan & Namasivayam, 1998). Methanol is toxic to the brain because elevated blood levels of methanol can cause severe changes in brain monoamine levels. It is well known that the nervous system is highly sensitive to methanol intoxication (Jeganathan & Namasivayam, 1998). Consumption of aspartame may also cause brain damage, depending on the high levels of phenylalanine in the blood (Haschemeyer & Haschemeyer, 1973).

2.3 Aspartame Usage Rules and Safety Assessment

The debate on the safety of aspartame has existed since its ratification in the 1980s. There are a large number of people and website organizations presenting a large number of anecdote literature detailing consumer concerns, which include cancer (Zehetner & McLean, 1999) multiple sclerosis, seizures, loss of memory, depression, blindness, anxiety, obesity, birth defects and death, With both acute and chronic exposure concerns (Camfield et al., 1992; Watts, 1991; Van den Eeden, et al., 1994; Lean & Hankey, 2004).

In a work of MRCA Information Service evaluated in the USA between the years of 1984-1992 aspartame consumption of about 2,000 households have been followed (Abrams, 1986; Butchko et al, 1996). In the 14-day-long questionnaire, all foods that individuals eat at home and outside were recorded and aspartame consumption by age groups was calculated in mg/kg. According to the results of the questionnaire, aspartame consumption was recorded as 1.6-3.0 mg/kg/day even in children, diabetics who were thought to be consuming aspartame the most. In addition, consumption of aspartame was gradually increased in the general population (1984-1985: 1.6 mg / kg / day, 1991-1992: 3.0 mg / kg / day). Another study of 1,500 women in the US regarding aspartame consumption reported that aspartame consumption of more than 90% of women was less than 5 mg / kg / day (Heybach and Smith, 1988).

Renwick's study of recent consumption of artificial sweeteners reported that aspartame consumption was 5.16 mg / kg / day for men and 4.64 mg / kg / day for women. In the general population, aspartame consumption in excess dose was found to be 13.29 mg / kg / day on average (Renwick, 2006).

A study conducted in Norway reports that the aspartame intakes of beverage consumers are below the ADI and range from 6.1 to 10.2 mg/kg bw, but the estimated intake might increase to levels above ADI if the contribution of other food categories is considered (Husøy et al., 2008). The daily intakes of aspartame by children and adolescents are estimated to be below the ADI (Butchko & Kotsonis, 1991).

When other works investigated in other countries, aspartame consumption per person in the general population in Australia was found to be 3 mg / kg / day (National Food Authority Australia, 1995). In a study of changes in nutrient intake, it was reported that Australians reduced consumption of artificial sweeteners significantly. However, the average daily consumption amount was not quantified (Rangan et al., 2011). In Brazil, it was reported that the general population consumed about 1.2 mg / kg / day, about 1.0 mg / kg / day for diabetic individuals and about 1.2 mg / kg / day for individuals who were on weight loss diet (Toledo and Ioshi, 1995). This amount was 1.1 mg / kg / day in Finland (Virtanen et al. 1988), 2.4 mg / kg / day in France (Garnier-Sagne et al. Biermann, 1992), 0.3 mg / kg / day in Italy (Leclercq et al. 1999), 1.9 mg / kg / day in the Netherlands (Hulshof and Bouman, 1992) and 0.9-3.4 mg / kg / day in Norway (Bergsten, 1993).

Aspartame usage has been regulated in more than 100 countries in accordance with health-based guidance values defined by the key regulatory authorities around the world, including the US Food and Drug Administration (FDA, 1984), and the Joint FAO/WHO Expert Committee on Food Additives (JECFA) (JECFA, 1980). In Europe, the following scientific organizations were responsible for advising on the safety of sweeteners: the Scientific Committee for Food (SCF from 1974 until April 2003), the EFSA Panel on Food Additives, Flavorings, Processing Aids and Materials in Contact with Food (from 2003 to 2008), and currently the EFSA Panel on Food Additives and Nutrient Sources Added to Food (ANS, from 2008 to present).

The first safety assessment guidelines for food additives were issued by SCF in 1980 (SCF, 1980). Later, SCF adopted its new guideline in 2001 (SCF, 2001). JECFA, which started its work in 1956, offers its results of evaluations to the member governments of the United Nations, in an informal manner. JECFA issued guidelines for safety assessment of food additives in 1987 (WHO, 1987).

These organizations and authorities have independently set different ADIs for aspartame of 50 mg/kg/day in the U.S. (FDA, 1984), 0–40 mg/kg bw (JECFA, 1980), and 0–40 mg/kg bw in European Union based on long-term rat studies using the NOAEL of 4 g/kg bw/day (SCF, 1985). Later on in 1988, 1997, 2002, SCF (SCF, 1989, 1997, 2002) and more recently EFSA (2005, 2006, 2009, 2011) performed additional reviews of aspartame data. Diketopiperazine (DKP), a minor cyclic dipeptide derivative of aspartame which is formed in some aqueous solutions, was also regulated by an ADI of 7.5 mg/kg bw/day (JECFA, 1980; SCF, 1985).

Food and beverages containing aspartame must include a label on their packaging indicating that the product contains phenylalanine. Individuals suffering from hereditary disease phenylketonuria, must strictly limit their intake of this amino acid; hence this label will be beneficial for people suffering from PKU disorder.

Contrary, normal and healthy individuals do not need to restrict their phenylalanine intake (Mortensen, 2006).

Despite all this, more than 90 countries (Health Canada, 2005), including the United States, Canada, countries in the European Union, Japan, Australia, and New Zealand has approved use of aspartame in food (Magnuson et al., 2007). As an example, the following foods are products which include aspartame and are widely available in the United States: breath mints, carbonated soft drinks, cereals, chewing gum, flavored syrups for coffee, flavored water products, frozen ice, frozen ice cream novelties, fruit spreads, sugar-free gelatin, hard candies, ice cream toppings, no-sugar-added or sugar-free ice creams, iced tea (powder and ready to drink), instant cocoa mix, jams and jellies, juice blends, juice drinks, maple syrups, meal replacements, mousse, no-sugar-added pies, noncarbonated diet soft drinks, nutritional bars, powdered soft drinks, protein nutritional drinks, puddings, candy chews, sugar-free chocolate syrup, sugar-free cookies, sugar-free ketchup, table-top sweeteners, vegetable drinks, and yogurt (drinkable, fat-free, and sugar-free) (Lim et al., 2006). In Turkey, aspartame (E951) is used in many products, such as reduced-energy or sugar-free non-alcoholic beverages, confections etc., sweeteners etc. It is necessary for the products containing aspartame to carry a label statement indicating that the product contains phenylalanine.

Despite aspartame intakes have increased since the 1980s, current estimates show that they all remain below the ADI level (Magnuson et al., 2007). A recent (2013) review of aspartame by the EFSA, based on 26 studies conducted in 17 European countries, estimated the mean exposure to be 1.2 to 5.3 mg/kg/d, while the highest consumption reaching 1.9 to 15.6 mg/kg/d (EFSA, 2013).

2.4 Biochemistry and Metabolism

Aspartame is solely consumed orally through intake of beverages, foods, and chewing gum, with a small amount used in oral pharmaceutical preparations (tablets and liquids) (Magnuson et al., 2007). Aspartame is the methyl ester of aspartyl dipeptide consisting of the amino acids L-aspartic acid and L-phenylalanine (Burgert et al., 1985). These three components are utilized in the body in the same way as when they are derived from foods, such as milk, fruit and vegetables. Further, these components are derived in much greater amount from common foods than beverages containing aspartame sweetener (Butchko & Stargel, 2001).

Aspartame is white, odorless, crystal dusty and sweet. It is almost completely soluble in water, partially soluble in alcohol (Sanyude and et al., 1991). It is expressed that the pH value at the isoelectric point 5.2, the melting point is 196°C and the deterioration temperature is 215°C (Maisons, 2002). The solubility of aspartame depends on temperature and pH. Aspartame is readily soluble in water at 25°C. It is very poorly soluble in methanol, not soluble in ethanol, chloroform and heptane. It is known that the solubility of aspartame at neutral pH is about half of its solubility at pH 3-4. The maximum solubility was observed at pH 2.2 (20 mg / ml at 25 ° C) and minimum solubility at pH 5.2 (pHi) (13.5 mg / ml at 25 ° C) (Maisons, 2002).

Orally ingested aspartame is absorbed from the small intestines and then metabolized by one of two ways. Aspartame is metabolized in the gastric tract to its three metabolites: aspartic acid (40%), phenylalanine (50%) and methanol (10%) (Stegink, 1987). Aspartame may be absorbed into the general circulation by completely being hydrolyzed to these three components in the gastrointestinal lumen, or may be hydrolyzed to methanol and aspartylphenylalanine dipeptide (Stegink,

1987). Aspartic acid is converts into alanine and oxaloacetate, phenylalanine converts into tyrosine and, to a lesser extent, phenylethylalanine, and methanol transforms into formaldehyde and then to formic acid (Ranney & Oppermann, 1979). Aspartic acid is a metabolite of aspartame and an excitatory amino acid which is normally found in high levels in the brain. The blood-brain barrier which protects the brain from large fluctuations in plasma aspartate controls these levels (Maher & Wurtman, 1987). Before aspartic acid (i.e. the carboxylate anion of aspartic acid) reaches the portal circulation and enters the free amino acid pool, in enterocytes it is transformed into oxaloacetate through transamination (Magnuson et al., 2007). Oxaloacetate and aspartate can participate in the urea cycle and gluconeogenesis by interconverting in the body (EFSA, 2013). Aspartate can also be useful for generating other essential amino acids (methionine, threonine, isoleucine, and lysine) and by stimulating the N-methyl-d-aspartate receptors, it can also functions as a neurotransmitter (Humphries, Pretorius, & Naude, 2007). Excess aspartate is excreted in the urine.

Oxaloacetate and aspartate can participate in the urea cycle and gluconeogenesis by interconverting in the body (EFSA, 2013). Aspartate can also be useful for generating other essential amino acids (methionine, threonine, isoleucine, and lysine) and by stimulating the N-methyl-d-aspartate receptors, it can also functions as a neurotransmitter (Humphries, Pretorius, & Naude, 2007). Excess aspartate is excreted in the urine.

Phenylalanine, which is found in almost all protein rich foods, is an essential amino acid to produce monoamine in the brain (Haschemeyer & Haschemeyer, 1973). After being partially converted to tyrosine by hepatic phenylalanine hydroxylase (Abdel-Salam et al., 2012) or to phenethylamine and phenylethanolamine, phenylalanine enters the plasma free amino acid pool from the portal blood. Phenylalanine entering the systemic circulation can be distributed throughout the body, including the brain, where it is needed for normal growth and development (Butchko et al. 2002).

Tyrosine is used in the biosynthesis of dopamine and norepinephrine. neurotransmittersPhenylalanine is highly concentrated in the human brain and plasma and is a precursor of the neurotransmitters called catecholamines, which are adrenalin-like substances. Phenylketonurics, people who have metabolic disorder in phenylalanine, have their serum plasma levels of phenylalanine raised up to 400 times normal (Chem, 2017). Phenethylamine is an internal non catecholamine similar to amphetamine in structure and in effects on spontaneous behavior (Mullenix et al. 1991).

Furthmore, the primary metabolic fate of methanol from aspartame goes into the portal circulation and then catalase-peroxidase (in rodents) or alcohol dehydrogenase (in primates and humans) convert it to formaldehyde and then into formate fastly. (Terphly, 1991). (Butchko, Stargel, Comer, & et al, Aspartame: review of safety., 2002). In a few minutes, formaldehyde is oxidized with formaldehyde dehydrogenase to formic acid, and the half-life of formaldehyde is between 1 and 2 minutes. Formic acid is removed from the body in two ways; it excreted from the body by the urine, or exhaled through the breath after metabolizing to carbon dioxide. Many animal and human studies have been researched the metabolism of methanol from aspartame, because of the adverse health effects of fast consuming of highe levels of methanol sufficient to cause methanol toxicity. When pathways of formic acid metabolism are suppressed methanol toxicity occurs and it causes an increase in the concentration in the body. The result of the wprks on humans have shown that in both cases with very large single doses of aspartame or repeated chronic exposure over time, there are no changes in baseline blood formate (the anion of formic acid) levels following consumption of aspartame-containing products (Magnuson et al. 2007). But the amount of the products which obtained from many other natural nutritional resources are much more than the amount of these digestion products (EFSA, 2013) (Butchko et al, 2002). for example, the amount of methanol in tomato juice is 6 times more than that derived from aspartame in diet cola (Butchko et al, 2002). Methanol is both freely present and is produced from other natural food ingredients such as pectin in the digestion of many foods and drinks, including fruits, juices, vegetables coffee, and alcoholic drinks (Magnuson et al. 2016). Phenylalanine and the amino acids aspartate (ie, anion of aspartic acid) are very common in the diet, found in foods such as dairy, lean protein, and beans (Butchko et al, 2002) .

The metabolism of aspartme are given in Figure 2.

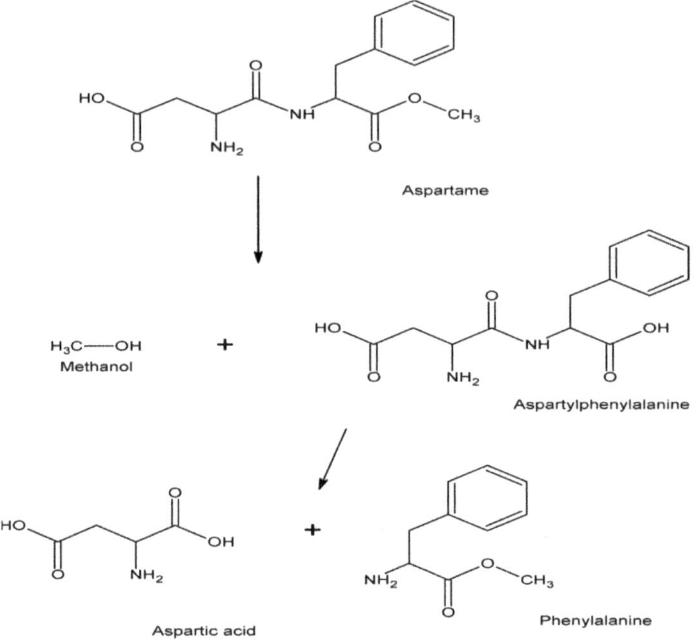

Fig 2. Metabolism of aspartame (Oppermann et al., 1973b; Ranney et al., 1976; Ranney and Oppermann, 1979; Oppermann and Ranney, 1979).

These compounds are intestinally absorbed and reach the portal blood in a similar way to amino acids, ethanol and dietary proteins and polysaccharides. On the other way, the aspartame lumen loses a methyl group to form aspartyl-phenylalanine and methanol. Aspartyl-phenylalanine is absorbed directly into mucosal cells by the peptide transport mechanism. In both cases, high doses of aspartame should be released into the portal blood aspartate, phenylalanine and methanol, and these compounds must be metabolized and / or eliminated from the body (Burgert and et al., 1985).

L-aspartic acid, which makes up 40% of the aspartame's structure, plays an important role in nitrogen and energy metabolism together with glutamic acid. It is locaded at the main entrance to the cycle of the tricarboxylic acid the functional energy-producing component of the cell. Aspartate enters the mitochondria and transaminates and forms oxaloacetate with a-ketoacids. If aspartic acid is present in the body, the excess aspartic acid is converted into fumarate entering the tricarboxylic acid cycle and takes part in energy generation (Stegink, 1984). It has been supposed that neither aspartame nor its metabolites accumulates in the body. The way of use these components in the body is just like when they were derived from common foods (Whitehouse et al. 2008).

2.5 Toxicological Profile

As in other artificial sweeteners, aspartame increases the consumption of nutrients without consuming added sugar and reduces energy intake. Consumption of aspartame is believed to be safe when it comes to the amount of use and ADI value of countries (EC, 2002;Renwick, 2006). However, aspartame has been a health concern since the date it was produced, studies on acute, chronic and toxic effects have been made.

Taking into account the metabolism of aspartame, toxicity studies have concentrated especially on methanol and phenylalanine in the aspartame structure. Methanol, a colorless, odorless and bitter alcohol that makes up 10% of aspartame, increases blood methanol and formic acid levels and causes a variety of adverse effects including metabolic acidosis and blindness. Tephly and McMartin's work has shown that the toxic effect is caused by the accumulation of formic acid from methanol and formaldehyde. Because methanol reaches the liver via portal blood, then is collapsed by alcohol dehydrogenase enzyme to formaldehyde, formaldehyde is destroyed by aldehyde dehydrogenase enzyme and toxic metabolite (Tephly and McMartin, 1984;Anderson , 2004).

Methanol poisoning is present in the clinic at a table between 40 minutes and 72 hours, with a silent period without symptoms. This symptomless period is caused by slow metabolism of formaldehyde by methanol. At the end of this period, vision disorders, headache, dizziness and confusion typically occur. Blood levels of methanol above 20 mg dL^{-1} are considered toxic. Blood methanol levels above 40 mg dL^{-1} lead to serious disorders, while over 80-100 mg dL^{-1} is lethal. Severe intoxication causes cerebral edema and coma, and convulsion is observed as a result of this (Liu et al., 1998).

L-phenylalanine and L-aspartic acid, which make up the structure of aspartame, are naturally present in many foods consumed daily in the human diet and are metabolized in the same way as amino acids present in meat, cheese, fish, vegetables, juices and mother's milk containing these amino acids.

When a food containing protein is digested, plasma phenylalanine levels increase and postprandial plasma phenylalanine levels rise to 9-12 μmol / dL. Severely elevated levels of plasma phenylalanine are associated with mental retardation for phenylketonuria (PKU) patients. In this genetic disorder, where phenylalanine hydroxylase is deficient, phenylalanine is unable to metabolize and accumulates in calm and tissues. Thus, the levels of plasma phenylalanine in children with PKU range from 120-600 μmol / dL (Koch and et al., 1974).

The main concern about phenylalanine, which makes up 50% of aspartame, is that phenylalanine in children with PKU leads to toxic values. In a study of newborn monkeys when plasma phenylalanine would be toxic, no problems were observed in the first 9 months of life when fed with diets containing 1,2 and 3 g / kg aspartame, even though plasma phenylalanine levels were increased (Suomi, 1984).

In normal individuals, phenylalanine, dietary phenylalanine, taken from aspartame and supplied from the amino acid pool, is transformed into tyrosine by the enzyme phenylalanine hydroxylase, while tyrosine is converted into fumarate and metabolized in the tricarboxylic acid cycle and is involved in energy generation. As an alternative, phenylalanine is converted into its metabolites by decarboxylation or transamination reactions and is excreted in the urine (Scriver and Kaufman, 2001).

The crystalline form of aspartame is rapidly degrading in aqueous media. Degradation with temperature and pH differences is additionally accelerating. Aspartame cyclizes in the form of diketopiperazine (3-phenylmethyl-2,5-diketopiperazine-6-acetic acid, DKP) or hydroxylation to the α-aspartyl phenylalanine (α-AP) form in which methanol is released. Thus, if the aspartame-flavored carbonated beverage is stored at 30 ° C for 8 weeks, 38% of the aspartame is degraded. 12% of the degradation products are DKP (Bell and Labuza, 1991). Although the DKP has been noted as a toxin at the beginning of the 1970s, it has been reported that even at very high doses it may detoxify in the liver and the ADI value is reported to be 7.5 mg / kg / day (Maisons, 2002).

In a study, the newborn was shown to be exposed to aspartame exposure used by pregnant mice, together with continued chronic exposure of the newborn to aspartame taken on a diet during the first 20 weeks of life, insulin sensitivity and deterioration of cognitive performance may lead to increased weight gain compared to controls, especially in men. The aspartame administration inside the ADI, did not affect the consumption of water and food (Collison et al 2012).

2.5.1 Acute Toxicity

Studies on the acute effect of aspartame have generally focused on the effects of aspartame metabolites on cognitive performance, behavioral change, sensory motor function related to learning and memory, catecholamine metabolism and large amino acid metabolism. In a study of cognitive performance, it was reported that 50 mg / kg aspartame had no adverse effect on cognitive performance regardless of other effects (Stokes and ark, 1991). Spontaneous behavior did not change significantly in rats given high doses of aspartame (500-1000 mg / kg), phenylalanine (281-562 mg / kg) or tyrosine (309-618 mg / kg) (Mullenix and ark, 1991). It was found that oral aspartame significantly changed behaviour, anti-oxidant status and morphology of the hippocampus in rats also, possibly it may trigger hippocampal adult neurogenesis (Onaolapo et al. 2017).

In a study evaluating high-dose aspartame intake of pre-school and school-age children, although aspartame consumption was higher in the pre-school period (38 ± 13 mg / kg) than in the school age (32 ± 8.9 mg / kg), there was no change in behavior and cognitive function in both groups (Wolraich, 1994). Randomized controlled double-blind another study in children showed that aspartame consumption of 34 mg / kg / day at 2 weeks did not affect behavior and cognitive status of children with attention deficit (Shaywitz, 1994). In a study of 10 healthy volunteers, the effect of single-dose aspartame (15 mg / kg) on cognitive functions, mood and duration of response was investigated and no effect was observed in cognitive functions, memory and response time, although there was an increase in levels of plasma phenylalanine (Lapierre et al., 1990).

In a study of the effects of aspartame (1000 mg / kg) and its metabolites on brain amines in two different rats, levels of norepinephrine, dopamine and serotinin in brain hypothalamus, cerebellum, pons, hippocampus, sitria, cortex and thalamus sections were examined and catecholamine and indolamine levels no significant reduction was observed (Freeman et al., 1990). It is known that even after a protein-containing diet, the concentration of phenylalanine increases, the concentration in the brain does not increase. This is because the large neutral amino acids (LNAs) are to be in competition with each other in the transport (Wurtman ve Maher, 1987). Increased levels of phenylalanine in the brain or decreased levels of tyrosine and tryptophan may result in decreased neurotransmitter concentrations. In another study in which aspartame aromatic amino acids were studied for their effect on the hydroxylation rates and levels of LNAA and aspartame (200 mg / kg) was administered parenterally, phenylalanine and tyrosine levels in the carotid and the brain were followed. Although the level of tryptophan in his brain decreased with time, it was found that the branched chain amino acid levels were not affected. No changes in serotonin, dopamine, norepinephrine, 5-hydroxyindole acetic acid, dihydroxy phenyl acetic acid and homovalinic acid levels were observed (Fernstrom and et al., 1983). Anxiety levels associated with aspartame in CD1 male rats was evaluated by the LaBuda et al. (2000). i.p. aspartame doses of vehicle, 1000, or 2000 mg/kg, were given to CD-1 male rats, followed 30 min later by i.p. ethanol doses of 1.6 g/kg or vehicle. The results indicated that the aspartame circumstance have no significant impact on anxiety-related demeanor and did it modify the anxiolytic effects of ethanol. Bergstrom and et al. have examined the effect of aspartame on the level of excited extracellular dopamine in the rat region of the sitria. They thought that a single dose (500 mg / kg) of aspartame might be a relatively potent effect in reducing the level of excited extracellular dopamine (Bergstrom and et al., 2007).

Impulsive aggression in rats is associated with the level of brain serotonin. Some studies have shown that concentrations of 5-hydroxyindole acetic acid, which is serotonin and metabolite, cause a decrease in cerebrospinal fluid levels (Shaw and et al., 1967;Stanley and Stanley, 1990). Goerss and et al. investigated the relationship between serotonin and aggression in aspartame-fed rats due to the phenylalanine involved in the synthesis of serotonin. 11 Long-Evans rats were given 200-800 mg / kg aspartame and the standard onset-aggressive paradigm test was performed. Test results showed that high-dose aspartame decreased aggressiveness due to serotonergic activity (Goerss, Wagner and Hill, 2000).

Schulpis and et al. found that 34 mg / kg aspartame in healthy adults had no effect on Na-K-ATPase activity, which is crucial for the maintenance of cell fluid and ion homeostasis in neurons, but 150 and 200 mg / kg aspartame use significantly reduced Na-K-ATPase activity (Schulpis, Papassotiriou, Parthimos, Tsakiris and Tsakiris, 2006). In a double-blind randomized cross-sectional study of aspartame susceptibility, aspartame did not cause any acute effects when compared with the group control group fed with 100 mg aspartame-containing snack food for at least 7 days (Sathyapalan and et al., 2015). A significant increase in lipid peroxidation levels of aspartame fed Wistar albino rats, glutathione peroxidase levels (GPx), superoxide dismutase activity (SOD), and catalase activity (CAT) with a important decrease in

GSH and protein thiol reported by Ashok et al. (2015). Aspartame exposure resulted in detectable methanol even after 24 hours. Authors have also concluded that methanol and its metabolites may be responsible for the generation of oxidative stress in brain regions.

Since the fetal period is considered to be vulnerable, maternal nutrition in this period has an important influence on fetal life. There are not many studies about aspartame used in this period. In another study evaluating the effect of aspartame used in the fetal period, 20 pregnant rats were divided into 4 groups and the effect of aspartame on maternal-fetal and placental weights, umbilical cord length and fetal liver length were investigated. Diluted aspartame was given to one of the groups at room temperature, while temperature was set at 40°C in the other group given diluted aspartame. For control groups, water was given at room temperature and the temperature was set at 40°C. At the end of the study, a decrease in placental and maternal-fetal weight and umbilical cord length was observed at room temperature and at 40 °C. As a result, it has been reported that the use of aspartame in pregnancy may be harmful to the fetus (Portela and et al., 2007).

In a study of Abdel-Salam et al. (2012) aspartame (0.625, 1.875 or 5.625 mg/kg) was applied for two weeks once daily subcutaneously and Swiss albino atsr were investigated four times a week for their ability to situate a submerged layer. In his brain, malondialdehyde (MDA), reduced glutathione(GSH), nitric oxide levels (the concentrations of nitrite/nitrate) and glucose were detected. The results showed that the aspartame dose of 5.625 mg / kg significantly degraded the water maze performance. After treatment with aspartame at 1.875 mg/kg, important increase in brain MDA by 16.5% and nitric oxide by 16.2% and a decrease in GSH by 25.1% and glucose by 22.5% occurred. There was significantly increased brain MDA by 43.8%, nitric oxide by 18.6% and decreased GSH by 32.7% and glucose by 25.8% by administering aspartam at 5.625 mg/kg.

İt was reported by Ashok et al. (2015) that there was a significant increase in lipid peroxidation levels of aspartame fed Wistar albino rats, superoxide dismutase activity (SOD), glutathione peroxidase levels (GPx), and catalase activity (CAT) with a significant reduction in GSH and protein thiol. Aspartame exposure resulted in detectable methanol even after 24 hours. The writers had this result that methanol and its metabolites may be responsible for the formation of oxidative stress in brain areas.

Energy intake, blood sugar and insulin reactions of NNS(non-nutritive sweeteners) and sucrose (65 g) was investigated. Authors used healthy male cases with four treatments: aspartame, monk fruit, stevia, and sucrose sweetened drinks. According to their result Ad libitum lunch consumption was importantly higher for the NNS treatments compared with sucrose. Also consumption of artificial and natural NNS sweetened calorie-free drinks has a minimum effect on total daily energy intake, postprandial glucose and insulin compared with a sucrose-sweetened drinks beverage (Tey et al. 2017).

In another study, it was reported that the association between body mass index (BMI) and glucose tolerance affected by aspartame intake, wherein only those reporting aspartame intake had a powerful positive association between BMI and glucose tolerance than those reporting no aspartame consumption. Aaspartame consumption is related with increasesed obesity-related disruptions in fasting glucose and glucose tolerance (Kuk & Brown, 2016). As we have seen, studies on the acute effects of aspartame did not reveal any adverse effects of aspartame administered at different doses, but more studies on the safety of aspartame use during pregnancy are needed. As 40 mg / kg aspartame, which is allowed to be taken daily, does not cause an acute effect, higher doses may be considered as a risk for acute effects.

2.5.2 Sub-Chronic Toxicity

Sufficient work has not been reported on the sub-chronic effect of aspartame. In a study summarizing subchronic effects, 6, 10 and 13 mg / kg aspartame was reported to have no adverse effects on dogs, rats and mice. In the same study, subchronic toxicity of DKP was examined and no effect was observed (Kotsonis and Hjelle, 1996). Due to limited work done and more up-to-date information on sub-chronic effects of aspartame, more study is needed.

2.5.3 Chronic Toxicity

Concerns about the chronic use of aspartame have focused on the cause of cancer, multiple sclerosis, blindness, stroke, memory loss, depression, anxiety, birth defects and death (Lean and Hankey, 2004;Camfield and et al., 1992). Aspartame has negative effects on many functions, especially in the brain. It is known to cause neurotoxicity particularly (Bergstrom, Gummings and Skaggs, 2007;Christian and et al., 2004). It has also been found that in some studies chronic aspartame consumption is caused by learning and memory impairment. Ion balance is thought to play an important role in spatial learning and memory control (Moseley and et al., 2007).

Although short-term animal studies of aspartame suggest that aspartame is safe, a recent study has shown that aspartame taken throughout life causes increased risk of lymphoma, leukemia, pelvis, ureter, and bladder cell carcinoma (Soffritti and et al.,

2006). Nevertheless, there was no correlation between increased risk of cancer in epidemiological studies on the health effects of aspartame consumption on humans (Schernhammer and et al., 2012). This is thought to be due to the use of retrospective methods designed with potential reminder and prejudiced selection and to work as a mistake in dietary assessment (Aune, 2012).

Studies on rats have revealed the risk of developing certain types of cancer with the use of aspartame. A group of researchers in Italy and France have investigated the types of cancer that develop in Sprague-Dawley rats exposed to certain amounts of aspartame throughout their fetal life. Researchers have confirmed an increase in malignant tumors in male rats exposed to aspartame during the fetal period. In addition, male and female rats fed 100 mg / kg / day aspartame were found to have an incidence of lymphoma and leukemia and an increase in the incidence of breast cancer especially in female rats. When compared to prenatal and postnatal period and lifetime period, it is reported that exposure to aspartame in prenatal period is directly related to risk of cancer (Soffritti and et al., 2007). Another research made by these researchers found an increase in aspartame doses used in renal pelvis and transitional cell carcinomas in the ureter and swanomas in peripheral nerves. It was emphasized that aspartame is a multipotential cancer agent even with these effects and below the allowable daily dose (20 mg / kg / day) (Soffritti and et al., 2006).

Since the beginning of aspartame use in 1981, it has been thought that there may be a relationship between aspartame and brain cancers, especially with the increase in glioblastomas (Soffritti and et al., 2006;Olney and et al., 1996). It is thought that the relationship between hematopoietic cancer and aspartame in rats may cause the same effect in humans. The National Center for Cancer Research The NIH-AARP

Diet and Health Study investigated the relationship between aspartame-containing beverage consumption and hematopoietic and brain cancers (Lim and e al., 2006). In this study, in which the nutritional status of the participants was assessed, no relation was found between the risk of brain cancers and consumption of aspartame.

In one study, Bosetti et al. investigated the relationship between artificial sweeteners and cancers of the stomach, pancreas and endometrium. Patients with 230 gastric cancers, 326 pancreatic cancers and 454 endometrial cancers who were referred to 3 hospitals in various parts of Italy were followed up for 2 years and consumption of aspartame, saccharin and other artificial sweeteners were analyzed by the food consumption frequency questionnaire. As a result of the obtained data, there was no relation between the use of artificial sweeteners and gastric, pancreatic and endometrial cancer risks. It has been found that there is a direct relationship between sugar consumption and the risk of stomach and pancreatic cancer inversely to consumption of artificial sweeteners in the same untreated data (Bosetti and et al., 2009). Similarly, the study of 3010 patients with bladder cancer who live in the ten geographical regions of the United States (USA) reported that the use of aspartame and other artificial sweeteners did not lead to an increased risk of bladder cancer (Hoover and Strasser, 1980).

In the study of long-term use of aspartame (180 days) in the Wistar rats investigated the effect of antioxidant defense state in liver, aminotransferase (ALT), aspartate aminotransferase (AST), alkaline phosphatase (ALT), gamma glutamyl transferase found a significant increase in their activities. On the other hand, glutathione caused a significant decrease in both the low dose (500 mg / kg / day) and the high dose group. Considering these results, long-term use of aspartame appears to damage the glutathione defense system and cause hepatocellular injury (Abhilash et al., 2011). Observation of these effects in the liver led to the suspicion that oxidative stress may also occur in the brain. In a study on albino rats, chronic aspartame use (75 mg / kg) and protein thiol levels, enzymes involved in the removal of free radicals, reduced glutathione, lipid peroxidation, and oxidative stress were examined. At the end of the study, a significant increase in catalase and glutathione peroxidase enzyme activity, a significant decrease in reduced glutathione and protein thiol levels were observed in lipid peroxidation, superoxide dismutase activity. Chronic aspartame use has been reported to cause oxidative stress in some parts of the brain, along with increases in metabolite levels and metabolites (Iyyaswamy and Rathinasamy, 2012). In another study, the same effects were observed in rats using aspartame for a long time (40 mg / kg / day) and it was emphasized that chronic effects of aspartame use should be considered (Ashok and Sheeladevi, 2015).

In a study of acute and chronic effects of aspartame on neuropsychological and neurophysiological functions, 48 volunteers were followed for about 6 months. Volunteers were divided into two groups and their aspartame use was set high (45 mg / kg / day) and low (15 mg / kg / day). Participants were tested every ten days and twenty days, and the tenth days were considered as acute effects and the twentieth days as chronic effects. At the end of the study, the plasma phenylalanine levels of the individuals were elevated for both groups and no effect on neuropsychological and neurophysiological and behavioral functions was observed even in acute and chronic use in high aspartame consuming group (Spiers et al., 1998).

There are not enough studies about the use of aspartame in childhood. Gurney and et al. investigated the effect of aspartame consumption by paying attention to the incidence of brain cancer in childhood. On the nineteenth west coast of the United States, the amount of aspartame consumed during pregnancy and lactation was assessed for children who had been diagnosed with brain tumors between 1984-1991 and their mothers. There was no increase in brain tumor risk in children with aspartame used during pregnancy and lactation at the end of the study (Gurney et al., 1997).

Choudhary et al (2016) investigated whether aspartame consumption for longer periods of time has any effect on the heart of Wistar albino rats. For this purpose, the animals were randomly divided into 4 groups of 6 animals (first group: control, second group: folate deficient diet fed animals, third group: the control animals which have been treated with aspatame, forth group: folate deficient diet fed animals treated with aspartame). Aspartame was given orally, dissolved in normal saline, for 90 days (40 mg/kg•bw/day). Animals fed with folate deficient diet were used to mimic human methanol metabolism, because hepatic folate content in humans is very low. Plasma corticosterone levels significantly increased by aspartame consumption and suggesting that aspartame may act as a chemical stressor. A significant increase in lipid peroxidation was detected, nitric oxide and protein carbonyl, and significant decrease in protein thiol, cardiac membrane bound ATPases (Na+, K+, Ca++, Mg++), enzymatic (SOD, CAT, GPX, G6PD, GR) and non-enzymatic antioxidants (GSH, Vit-C, Vit-E) significant increase in both heart rate and heart marker enzymes (CK and CK-MB). It may be because of excessive generation of free radicals, which disrupts cardiac function. According to their result, the causal factors behind these changes, may be aspartame metabolite methanol or formaldehyde.

There are some studies about aspartame consumption and weight loss. In a study which assumed aspartame consumption may help to the development of metabolic syndrome according to phenylalanine inhibition of endogenous intestinal alkaline phosphatase (IAP). They used in vitro model, IAP was added to diet and regular soda, and IAP activity was measured. They used in vitro model, IAP was added to diet and regular soda, and IAP activity was quantified. For the acute model, they made closed bowel loop in rats. It was inoculated by aspartame or water and IAP activity was measured. Rat were given feed or high-fat diets (HFD) with/without aspartame whithin the drinking water for 18 weeks, for the chronic model. The results of in vitro studies were IAP activity is low when compared with controls in aspartame-containing solutions. Endogenous IAP activity in the aspartame group decreased by 50% compared with controls, for the acute model. Mice in the HFD + aspartame group gained more weight than the HFD + water group, for the chronic model. İmportant difference in glucose intolerance between the HFD ± ASP groups. Fasting glucose and serum tumor necrosis factor-alpha levels rose significantly in the HFD + aspartame group. They finalized that, the protective effects of endogenous IAP on metabolic syndrome can be inhibited by phenylalanine., a metabolite of aspartame, maybe it clarify the lack of expected weight loss, and metabolic improvements associated with diet drinks (Gul, et al., 2017).

Also a study indicated that neither aspartame nor its digestion products ever reach the colon; therefore, aspartame itself cannot directly affect intestinal microbiota.3 With this information in mind, it is important to carefully investigate the study design and other parameters that may be responsible for the differences in intestinal microflora observed in studies of aspartame-fed animals (Magnuson, Carakostas, Moore, Poulos, & Renwick, 2016). Palmnäs et al. described the changes in the intestinal microflora in rats fed either a low-fat (12%) or a high-fat (60%) diet and given either plain water or the water which sweetenered with aspartam. Especially, mice given water that contain aspartam consumed 17% to 25% less calories from consumption of their diets, resulting in significantly less fat, fiber, protein, and other nutrients, which are familiar to modify intestinal microflora (Palmnas et al. 2014) .

Studies on the chronic effects of aspartame have shown some conflicting results in studies of animals and humans for some cancer types of aspartame administered at different doses. Chronic consumption has been observed to have various negative effects on health. In order to evaluate the safety of aspartame for long-term use, different doses and durations of study are needed.

2.5.4 Genotoxicity

Genotoxicity is a chemical component that causes damage to the genetic construct in cells due to mutation. Some of the studies on aspartame show that it may cause genotoxicity, while others define it as non-genotoxic (Rencüzogullari et al., 2004).

AlSuhaibani (2010) searched the safety of aspartame and metabolic degradation products (phenylalanine, aspartic acid and methanol) in vivo using chromosomal deviation (CAE) testing and sister chromatid Exchange (SCE) testing in rat bone marrow cells. Swiss Albino male rats were exposed to aspartame doses of 3.5, 35, 350 mg / kg body weight. Yazarlar, kardeş kromatid değişimine neden olmamasına rağmen aspartam tedavisinin tüm konsantrasyonlarda doza bağımlı kromozomal sapmalara neden olduğunu bildirmiştir. On the other hand, aspartame did not reduce mitotic index (MI).

Mukopadhyay et al. studied the chromosomal degradation of bone marrow cells isolated from the femur to evaluate effect of different doses of aspartame (3.5, 35, 350 mg / kg) on Swiss Albino female rats. At the end of 18 weeks no genotoxic effect was observed in aspartame alone or in combination with acesulfame-K (Mukhopadhyay et al., 2000).

Blood samples taken from 4 healthy individuals were exposed to 500, 1000 and 2000 µg / ml aspartame for 24 and 48 hours during the study conducted by Rencuzoğulları and et al. As a result of the study, it was found that aspartame causes chromosomal damage and micronucleus formation and cytotoxic effect by decreasing mitotic index (Rencüzogullari et al., 2004).

Sasaki et al. examined the effects of chemicals on coloration and preservatives, antioxidants, fungicides and sweeteners in the stomach, colon, liver, lung, brain, bone marrow, kidneys and stomach of DDY mice. It was reported that cyclamate and sucralose caused DNA damage in various organs, with aspartame at 2000 mg / kg / day reported not to cause DNA damage in any organ (Sasaki et al., 2002). In a further study of the genotoxicity of aspartame and other artificial sweeteners, aspartame, 150, 300, and 600 mg / kg acesulfame-K and 50, 100 and 200 mg / kg saccharin at 7, 14, 21, 28 and 35 mg the effects on DNA damage in bone marrow were examined. As a result of the study, it has been reported that acesulfame-K and saccharin cause more DNA damage than aspartame in comet parameters, and no mutagenic effect in Ames / Salmonella / microsome test (Bandyopadhyay et al., 2008).

Another subject that has been investigated for the relationship between aspartame and genotoxicity is the effect on ochratoxin, which is called mycotoxin, from human food all over the world and genetic damage. Genotoxicity of ocratoxin A has been reported in a study (Creppy et al., 1985). In a study by Creppy et al. 25 mg / kg / 48 h aspartame given in combination with ochratoxin A for several weeks prevented the genotoxic effect of ochratoxin A (Creppy et al., 1996).

When the results of these studies are evaluated, it is considered that the results of the studies are controversial because studies on the genotoxic effect of aspartame do not contain definitive information. New study is needed to provide more accurate information.

2.6 Other Studies Related to Aspartame

Much of the other work on aspartame has focused on the effect of aspartame on diseases in which the increase in the prevalence of obesity due to excessive energy intake and the increase in the intake of artificial sweeteners is due to a restriction on energy intake, which is the reason for primer usage.

Gallus et al. (2007) assessed a case control study to examine the role of sweeteners on cancer risk. Between 1991 and 2004, an integrated network of case-control studies was established in Italy. Cases were 598 patients with event, histologically confirmed cancers of the oral cavity and pharynx, 1225 of the colon, 304 of the oesophagus, 460 of the larynx, 728 of the rectum, 1031 of the ovary, 2569 of the breast, 1294 of the prostate and 767 of the kidney (renal cell carcinoma). Rate ratios for consumption of other sweeteners, mainly aspartame, were 0.77 (95% CI 0.39–1.53) for cancers of the oral cavity and pharynx, 0.80 (95% CI 0.65–0.97) for breast, 0.77 (95% CI 0.34–1.75) for oesophageal, 0.71 (95% CI 0.50–1.02) for rectal, 0.90 (95% CI 0.70–1.16) for colon, 0.75 (95% CI 0.56–1.00) for ovarian, 1.62 (95%

CI 0.84–3.14) for laryngeal, 1.23 (95% CI 0.86–1.76) for prostate and 1.03 (95% CI 0.73–1.46) for kidney cancer. The authors state that the studies they conducted showes lack of relationship between saccharin, aspartame, and other sweeteners, and a few common neoplasia risks.

Aspartame is important in adipogenesis by suppressing the activities of phosphorylated peroxisome proliferator-activating receptor γ (p-PPARγ), peroxisome proliferator-activating receptor γ (PPARγ), fatty acid binding protein 4 (FABP4) and sterol regulatory element-binding protein 1 (SREBP1) Effect on 3T3-L1 differentiation thought to play a role. 3T3-L1 fat cells were left to differentiate with 10 μg / ml aspartame for 6 days and the results were analyzed with the Oil Red O solution used for the determination of neutral oils. The results of the analysis showed significant reduction of PPARγ, FABP4 and SREBP1 samples of adipocyte treatment containing aspartame. These results have shown that aspartame may be a potential effect in the control of obesity (Pandurangan and et al., 2014).

Anton and et al. examined the effects of aspartame, stevia and sucrose on nutrient uptake, postprandial glucose and insulin levels. When 19 healthy, weak individuals aged 18-50 and 12 obese individuals were fed with meals containing aspartame, stevia and sucrose at lunch and dinner, they were followed by the starvation and toughness levels before evening meal. As a result of individual Beck depression scale, food census, eating disorder diagnosis scale and statistical analysis, stevian showed lower levels of postprandial glucose and insulin levels compared to aspartame and sucrose (Anton and et al., 2010). In another study, it was reported that aspartame contributes more to energy intake and weight gain relative to sucrose (Feijó and et al., 2013).

A recent study by Palmas et al. (2014) investigated the effect of low-dose aspartame on microbiota in diabetic obese rats with low fat (12% / total energy) and high fat (60% Day) aspartame in 8 weeks. Although aspartame consumes less energy and has less weight gain, it has been found that aspartame increases fasting blood sugar and has a negative effect on insulin tolerance test. When the bacterial composition was analyzed by faecal analysis, it was observed that there was an increase in *Enterobacteriaceae and Clostridium* leptin species. Analysis of serum metabolomics reported that aspartame was rapidly metabolized in the small intestine lumen and was associated with an increase in short chain fatty acids, which is thought to be a negative effect on insulin tolerance, leading to a high glyconeogenic effect (Palmnas and et al., 2014). Another study that tackles the relationship between aspartame and other artificial sweetener-containing beverage consumption and diabetes has been to follow the blood glucose homeostasis included in the study of young diabetic individuals living in Brazil. The results of this study show that individuals with normal weight use of artificial sweeteners and those with higher prevalence of diabetes have higher rates of fasting blood glucose and lower beta cell function (Yarmolinsky and et al., 2015).

3. CONCLUSION

In addition to the developments in the food market, consumers are also presented with sugar-free products containing sweeteners, consumers have more information on food safety issues and want more information about the safety of food additives. Therefore, the risk assessment process in the EU faces more transparency requests. The EFSA website (http:// efsa.europa.eu) is an example of the inclusion of the transparency principle in the risk assessment of chemical substances in the European foodstuff. All sweeteners permitted to be consumed in the EU have been subjected to a comprehensive safety assessment before being accepted. Their safety has been documented as a consequence of many in vitro and in vivo animal studies, tests in people, and in some cases, epidemiological studies. Thus, consumption of sweeteners in the quantities within ADI is not a health hazard for consumers.

4. REFERENCES

Abdel-Salam, O., Salem, N., El-Shamarka, M., Hussein, J., Ahmed, N., & El-Nagar, M. (2012). Studies on the effects of aspartame on memory and oxidative stress in brain of mice. *European Review for Medical and Pharmacological Sciences, 16*, 2092-2101.

Abhilash, M., Sauganth Paul, M. V., Varghese, M. V., & Harikumaran Nair, R. (2011). Effect of long term intake of aspartame on antioxidant defense status in liver. *Food and Chemical Toxicology, 49*, 1203-1207.

Abrams, I. J. (1986). Using the menu census survey to estimate dietary intake: :Postmarket surveillance of aspartame. *In International Aspartame Workshop Proceedings.* Washington.

Administration, F. a. (1980). DEPARTMENT OF HEALTH AND HUMAN SERVICES. *Federal Register, 45*(181), 61475-61481.

Anderson , I. B. (2004). *Methanol. In: Poisoning and Drug Overdose.* UD.

Anton, S. D., Martin, C. K., Han, H., Coulon, S., Cefalu, W. T., Geiselman, P., & Williamson, D. A. (2010). Effects of stevia, aspartame, and sucrose on food intake, satiety, and postprandial glucose and insulin levels. *Appetite, 55*, 37-43.

Ashok, I., & Sheeladevi, R. (2015). Oxidant stress evoked damages in rat hepatocyte leading to triggered Nitric oxide synthase (NOS) levels on long term consumption of aspartame. *journal of food and drug analysis, xxx*, 1-13.

Aune, D. (2012). Soft drinks, aspartame, and the risk of cancer and cardiovascular disease. *Am J Clin Nutr, 96*(6), 1249-1251.

Bandyopadhyay, A., Ghoshal, S., & Mukherjee, A. (2008). Genotoxicity Testing of Low-Calorie Sweeteners: Aspartame, Acesulfame-K, and Saccharin. *Drug and Chemical Toxicology, 31*, 447-457.

Bar, A., & Biermann, C. (1992). Intake of intense sweeteners in Germany. *Z. Ernahrungswiss, 31*, 25-39.

Bell, L. N., & Labuza, T. P. (1991). Aspartame degredation as a function of "water activity". *Adv Exp Med Biol, 302*, 337-349.

Bergamo, A., Alberto Fracassi da Silva, J., & Pereira de Jesus, D. (2011). Simultaneous determination of aspartame, cyclamate, saccharin and acesulfame-K in soft drinks and tabletop sweetener formulations by capillary electrophoresis with capacitively coupled contactless conductivity detection. *Food Chemistry, 124*, 1714-1717.

Bergsten, C. (1993). *Intake of Acesulfame-K, Aspartame, Cyclamate and Saccharin in Norway.* Oslo: Norwegian Food Control Authority.

Bergstrom, B. P., Cummings, D. R., & Skaggs, T. A. (2007). Aspartame decreases evoked extracellular dopamine levels in the rat brain: An in vivo voltammetry study. *Neuropharmacology, 53*(8), 967-974.

Bergstrom, B. P., Gummings, D. R., & Skaggs, T. A. (2007). Aspartame decreases evoked extracellular dopamine levels in the rat brain: an in vivo voltammetry study. *Neuropharmacology, 53*, 967-974.

Blackburn, G. L., Kanders, B. S., Lavin , P. T., Keller, S. D., & Whatley, J. (1997). The effect of

    aspartame as part of a multidisciplinary weight control program on short-and long

    term control of body weight. *American Journal of Clinical Nutrition , 65*(2), 409-418.

Blau, N., van Spronsen, F.J., Levy, H.L . (2010) Phenylketonuria. *Lancet* 76;9750), 1417–

1427.

Bopp, B. A., Sonders, R. C., & Kesterson, J. W. (1986). Toxicological aspects of cyclamate and

    cyclohexylamine. *Crit Rev Toxicol, 16*, 213-306.

Bosetti, C., Gallus, S., Talamini, R., Montella, M., Franceschi, S., Negri, E., & La Vecchia, C.

    (2009). Artificial Sweeteners and the Risk of Gastric, Pancreatic, and Endometrial

    Cancers in Italy. *Cancer Epidemiol Biomarkers Prev, 18*(8), 2235-2238.

Burgert, S. L., Merrick, R. H., Coon , J. D., Takeuchi, H., & Stegink, L. (1985). Intestinal

    metabolism of aspartame and its diketopiperazine and Lphenylalanine methyl ester

    degradation products. *American Journal of Clinical Nutrition, 41*, 867.

Butchko, H. H., Tshanz, C., & Kotsonis, F. N. (1996). Postmarketing surveillance of anecdotal

    medical complaints. *In The Clinical Evaluation of A Food Additive: Assessment of

    Aspartam* (s. 183-193). içinde Boca Raton: CRC Press.

Butchko, H., & Kotsonis, F. (1991). Acceptable daily intake vs actual intake: the aspartame

    example. *J. Am. Coll. Nutr, 10*, 258-266.

Butchko, H., & Stargel, W. (2001). Aspartame: scientific evaluation in the post marketing

    period. *Regul Toxicol Pharmacol, 34*, 221-233.

Butchko, H., Stargel, W., Comer, C., & et al. (2002). Aspartame: review of safety. *Regul

    Toxicol Pharmacol, 35*, S1-S93.

Butchko, H., Stargel, W., Comer, C., Mayhew, D., Beninger, C., Blackburn, G., . . . Trefz, F. (2002). Intake of aspartame versus the Acceptable Daily Intake. *Regul. Toxicol. Pharmacol, 35*, S13-S16. doi:10.1016/j.fct.2013.07.040

Camfield, P. R., Camfield, C. S., Dooley, J. M., Gordon, K., & Jollymore, S. (1992). Aspartame exacerbates EEG spike-wave discharge in children with generalized absence epilepsy: a double-blind controlled study. *Neurology, 42*, 1000-1003.

Cheftel, J. C. (2005). Analytical, Nutritional and Clinical Methods Food and nutrition labelling in the European Union. *Food Chemistry, 93*, 531-550.

Chem, O. (2017, 04 10). *L-Phenylalanine*. U.S. National Library of Medicine: https://pubchem.ncbi.nlm.nih.gov/compound/L-phenylalanine#section=Top adresinden alındı

Choudhary, A., Sundareswaran, L., & Sheela Devi, R. (2016). Aspartame induced cardiac oxidative stress in Wistar albino rats. *Nutrition Clinique et métabolisme, 30*, 29-37.

Christian, B., McConnaughey, K., Bethea, E., Brantle, S., Goffey, A., & Hammond, L. (2004). Chronic aspartame affects-maze performance, brain cholinergic receptors and Na,K-ATPase in rats. *Pharmacol Biochem Behav, 78*, 121-127.

Collison, K.S., Makhoul, N.J., Zaidi, M.Z., Saleh, S.M.i Andres, B., et al. (2012) Gender Dimorphism in Aspartame-Induced Impairment of Spatial Cognition and Insulin Sensivity. PLoS ONE 7(4), e31570. doi:10.1371/journal.pone.0031570

Creppy , E. E., Baudrimont, I., Belmadani , A., & Betbeder , A. M. (1996). ASPARTAME AS A PREVENTIVE AGENT OF CHRONIC TOXIC EFFECTS OF OCHRATOXIN A IN EXPERIMENTAL ANIMALS. *J. TOXICOL.-TOXIN REVIEWS, 15*(3), 207-221.

Creppy, E. E., Kane, A., Dirheimer, G., Lafarge-Frayssinet, C., Mousset, S., & Frayssinet, C. (1985). Genotoxicity of ochratoxin A in mice : DNA single-strand breaks evaluation in spleen liver and kidney. *Toxicology Letters, 28*(1), 29-35.

Dailey, J., Lasley , S., Burger , R., & et al. (1991). Amino acids,monoamines and audiogenic seizures in genetically epilepsy-prone rats: effects of aspartame. *J Epilepsy Res, 8*(2), 122-133.

Diomede, L., Romano, M., Guisio, G., & et al. (1991). Interspecies and interstrain studies on the increased susceptibility to metrazol-induced convulsions in animals given aspartame. *Food Chem Toxicol, 29*(2), 101-106.

EC. (2002). *Opinion of the Scientific Committee on Food: update on the safety of aspartame.* Brussels, Belgium: European Commission, Health and Consumer Protection Directorate-General.

EFSA, 2005. Opinion of the Scientific Committee on a request from EFSA related to a harmonised approach for risk assessment of substances which are both genotoxic and carcinogenic. EFSA J. 282, 1–31.

EFSA, 2006. Opinion of the scientific panel AFC related to new long-term carcinogenicity study on aspartame. EFSA J. 356, 1–44.

EFSA, 2009. Guidance of the Scientific Committee on a request from EFSA on the use of the benchmark dose approach in risk assessment. EFSA J. 1150, 1–72.

EFSA, 2011a. Statement on two recent scientific articles on the safety of artificial sweeteners. EFSA J. 9, 1996.

EFSA, 2011b. EFSA Statement on the scientific evaluation of two studies related to the safety of artificial sweeteners. EFSA J. 9, 2089.

EFSA, 2011c. Overview of the procedures currently used at EFSA for the assessment of dietary exposure to different chemical substances. EFSA J. 9 (12), 2490.

EFSA. (2013). Scientific Opinion on the re-evaluation of aspartame (E 951) as a food additive. *EFSA J, 11*, 3496. doi:10.2903/j.efsa.2013.3496.

FDA, 1984. Food additives permitted for direct addition to food for human consumption: aspartame. Food and Drug Administration. Federal Register 49, 6672.

FDA. (2006). *Artificial sweeteners: No calories...sweet!* www.fda.gov/fdac/features/2006/406_sweeteners.html.

FDA. (2007). *FDA Statement on European Aspartame Study.* Silver Spring MD.

Feijó, F. M., Ballard, C. S., Foletto, K. C., Melo Batista, B. A., Magagnin Neves, A., Marques Ribeiro, M. F., & Bertoluci, M. C. (2013). Saccharin and aspartame, compared with sucrose, induce greater weight gain in adult Wistar rats, at similar total caloric intake levels. *Appetite, 60,* 203-207.

Fernstrom, J. D., Fernstrom, M. H., & Gillis, M. A. (1983). Acute effects of aspartame on large neutral amino acids and monoamines in rat brain. *Life Sci, 32*(14), 1651-1658.

Freeman, G., Sobotka, T., & Hattan, D. (1990). Acute effects of aspartame on concentrations of brain amines and their metabolites in selected brain regions of Fischer 344 and Sprague-Dawley rats. *Drug Chem Toxicol, 13*(2-3), 113-133.

Fry, J. (1999). The world market for intense sweeteners. (A. Corti, Dü.) *World Rev Nutr Diet, 85*, 201-211.

Furia, T. E. (1972). *CRC Handbook of food addititives* (Cilt 2.). CRC press.

Garnier-Sagne, I., Leblanc, J. C., & Verger. (2001). Calculation of the intake of three intense

sweeteners in young insulin-dependent diabetics. *Food Chem. Toxicol, 39*, 745-749.

Goerss, A. L., Wagner, G. C., & Hill, W. L. (2000). Acute effects of aspartame on aggression

and neurochemistry of rats. *Life Sciences, 67*, 1325-1329.

Gordon, G. (1987). NutraSweet: questions swirl. UPI Investigative Report. *Reprinted in U.S.

Senate*, 483-510.

Graves, & Florence. (1985). How Safe is Your Diet Soft Drink. *Common Cause Magazine.*

Griffin, R.F., Humienny, M.E., Hall, E.C., Elsas, L.J. (1973) Classic phenylketonuria:

heterozygote detection during pregnancy. Am J Hum Genet 25(6), 646–654.

Gul, S., Hamilton, R., Munoz, A., Phupitakphol, T., Liu, W., Hyoju, S., . . . Hodin, R. (2017).

Inhibition of the gut enzyme intestinal alkaline phosphatase may explain how

aspartame promotes glucose intolerance and obesity in mice. *Appl. Physiol. Nutr.

Metab., 42*, 77-83.

Gurney, J. G., Pogoda, J. M., Holly, E. A., Hecht, S. S., & Preston-Martin, S. (1997).

Aspartame Consumption in Relation to Childhood Brain Tumor Risk: Results From a

Case–Control Study. *Journal of the National Cancer Institute, 89*(14), 1072-1074.

Haschemeyer, R., & Haschemeyer, A. (1973). *Proteins. In: A Guide to Study by Physical and

Chemical Methods.* New York: John Wiley & Sons.

Heybach, J. P., & Smith, J. L. (1988). *). Intake of aspartame in 19–50 year old women from

the USDA continuing survey of food intakes by individuals (CSFII 85).* A1615: FASEB

J.

Hoover, R., & Strasser, P. H. (1980). ARTIFICIAL SWEETENERS AND HUMAN BLADDER

CANCER: Preliminary Results. *The Lancet, 315*, 837-841.

Hulshof, K. F., & Bouman, M. (1992). *Use of Various Types of Sweeteners in Different

Population Groups.* Netherlands.: TNO Nutrition and Food Research Institute.

Humphries, P., Pretorius, E., & Naude, H. (2007). Direct and indirect cellular effects of

aspartame on the brain. *Eur. J. Clin. Nutr, 62*, 451-462.

Husøy, T., Mangschou, B., Fotland, T., Kolset, S., Nøtvik Jakobsen, H., Tømmerberg, I., . . .

Frost Andersen, L. (2008). Reducing added sugar intake in Norway by replacing

sugar sweetened beverages with beverages containing intense sweeteners – a risk

benefit assessment. *Food Chem. Toxicol, 46*, 3099-3105.

Iyyaswamy, A., & Rathinasamy, S. (2012). Effect of chronic exposure to aspartame on

oxidative stress in brain discrete regions of albino rats. *J. Biosci, 37*(4), 679-688.

Jacohsen, D., & McMartin, K. E. (1986). Methanol and ethylene glycol poisonings.

Mechanism of toxicity, clinical course, diagnosis and tratment. *Med Toxicol, 1*, 309-

334.

Jagenburg, R., Rega°rdh, C.G., Ro¨djer, S. (1977) Detection of heterozygotes for

phenylketonuria. Total body phenylalanine clearance and concentrations of

phenylalanine and tyrosine in the plasms of fasting subjects compared. *Clin

Chem* 23(9), 1654–1660.

JECFA, 1980. Aspartame; Evaluation of certain food additives. Joint FAO/WHO

Expert Committee on Food Additives. Technical Report Series 653. World Health

Organization, Geneva.

Jeganathan, P., & Namasivayam, A. (1998). Methanol induced biogenic amine changes in discrete areas of rat brain: Role of simultaneous ethanol administration. *J Ind J Physiol Pharmacol, 32*(1), 1-10.

Jeganathan, P., & Namasivayam, A. (1998). Methanol induced biogenic amine changes in discrete areas of rat brain: Role of simultaneousethanol administration. *Ind J Physiol Pharmacol, 32*, 1-10.

Kalkhoff, R., & LEVIN, M. (1978). The saccharin controversy. *Diabetes Care, 1*, 211.

Koch, R., Blaskowics, M., Wenz , E., Fishler, K., & Schaeffler, G. (1974). *Phenylalaninemia and phenylketonuria.* (W. L. Nyhan, Dü.) New York.

Kotsonis, F. N., & Hjelle, J. J. (1996). In The Clinical Evaluation of a Food Aditive Assessment of Aspartame. C. Tschanz, H. H. Butchko, W. W. Stargel, & F. N. Kotsonis (Dü) içinde, *The safety assessment of aspartame: Scientific and regulatory considerations.* (s. 23-41). Boca Raton: CRC Press.

Kruse, J. A. (1992). Methanol poisoning. *Intensive Care Med, 18*(7), 391-397.

Kuk, J., & Brown, R. (2016). Aspartame intake is associated with greater glucose intolerance in individuals with obesity. *Appl. Physiol. Nutr. Metab., 41*, 795-798.

Lapierre, K. A., Greenblatt, D. J., Goddard, J. E., Harmatz, J. S., & Shader, R. I. (1990). The neuropsychiatric effects of aspartame in normal volunteers. *Journal of Clinical Pharmacology, 30*, 454-460.

Lean, M. E., & Hankey, C. R. (2004). Aspartame and its effects on health. *BMJ, 329*, 755-756.

Leclercq, C., Berardi, D., Sorbillo, M. R., & Lambe, J. (1999). Intake of saccharin, aspartame,

acesulfame K and cyclamate in Italian teenagers:Present levels and projections. *s.*

Food Addit. Contam, 16, 99-109.

Lim, U., Subar, A. F., Mouw, T., Hartge, P., Morton, L. M., Stolzenberg-Solomon, R., . . .

Hollenbeck, A. R. (2006). Consumption of Aspartame-Containing Beverages and

Incidence of Hematopoietic and Brain Malignancies. *Cancer Epidemiol Biomarkers*

Prev, 15(9), 1654-1659.

Liu, J. J., Daya, M. R., Carrasquillo, O., & Kales, S. N. (1998). Prognostic factors in patients

with methanol poisoning. *Japonesse Toxicology Clinical Toxicology, 36*, 175-181.

Magnuson, B., Burdock, G., Doull, J., Kroes, R., Marsh, G., Pariza, M., . . . Williams, G. (2007).

Aspartame: A Safety Evaluation Based on Current Use Levels, Regulations, and

Toxicological and Epidemiological Studies. *Critical Reviews in Toxicology, 37*, 629-

727.

Magnuson, B., Carakostas, M., Moore, N., Poulos, S., & Renwick, A. (2016). Biological fate of

low-calorie sweeteners. *Nutrition reviews, 11*, 670-689.

Maher, T., & Wurtman, R. (1987). Possible neurologic effects of aspartame, a widely used

food additive. *J Environ Health Persp, 75*, 53-57.

Maisons, A. (2002). *Opinion on a possible link between the exposition to aspartame and the*

incidence of brain tumours in humans . Afssa.

Market Evaluation Consumption and Statistics Committee (MECAS). (2012). *Alternative*

Sweeteners in a Higher Sugar Price Environment. London: International Sugar

Organization. Nisan 2012 tarihinde alındı

Mazur, R. (1984). Discovery of aspartame. M. R. H, L. Stegink, & L. Filer (Dü) içinde, *Aspartame: Physiology and biochemistry* (s. 3-9). New York: Marcel Dekker.

Mazur, R. H., Schlatter , J. M., & Goldkamp, A. H. (1969). Structure-Taste Relationships of Some Dipeptides. *Journal of the American Chemical Society, 91*(10), 2684-2691.

Mortensen, A. (2006). Sweeteners permitted in the European. *Scandinavian Journal of Food and Nutrition, 50*(3), 104-116.

Moseley, A. E., Williams, M. T., & Schaefer, T. L. (2007). Deficiency in NaþKþ-ATPase alpha isoform genes alters spatial learning, motor activity, and anxiety in mice. *J Neurosci, 27*(3), 616-626.

MUKHOPADHYAY, M., MUKHERJEE, A., & CHAKRABARTI, J. (2000). In Vivo Cytogenetic Studies on Blends of Aspartame and Acesulfame-K. *Food and Chemical Toxicology , 38*, 75-77.

Mullenix, P. J., Tassinari, M. S., & Kernan, W. J. (1991). No change in spontaneous behavior of rats after acute oral doses of aspartame, phenylalanine, and tyrosine. *Fundam Appl Toxicol, 16*(3), 495-505.

National Food Authority Australia. (1995). *Survey of Intense Sweetener Consumption in Australia.* Canberra: National Food Authority.

Negro, F., Mondardini, A., & Palmas, F. (1994). Hepatotoxicity of saccharin. *New England Journal of Medicine, 331*, 134-135.

Olney, J. W., Farber, N. B., Spitznagel, E., & Robins, L. N. (1996). Increasing brain tumor rates: is there a link to aspartame? *J Neuropathol Exp Neurol, 55*, 1115-1123.

Onaolapo, A., Onaolapo, O., & Nwoha, P. (2017). Aspartame and the hippocampus:

Revealing a bi-directional, dose/time-dependent behavioural and morphological

shift in mice. *Neurobiology of Learning and Memory, 139*, 76-88.

Oyama, Y., Sakai, H., Arata, T., Okano, Y., Akaike, N., Sakai, K., & Noda, K. (2002). Cytotoxic

effects of methanol, formaldehyde, and formate on dissociated rat thymocytes: a

possibility of aspartame toxicity. *Cell Biol Toxicol, 18*, 43-50.

Palmnas, M. S., Cowan, T. E., Bomhof, M. R., Su, J., Reimer, R. A., Hogel, H. J., . . . ShearerJ.

(2014). Low-Dose Aspartame Consumption Differentially Affects Gut Microbiota-

Host Metabolic Interactions in the Diet Induced Obese Rat. *Plos One, 9*(10),

e109841.

Pandurangan, M., Park, J., & Kim, E. (2014). Aspartame downregulates 3T3-L1

differentiation. *In Vitro Cell.Dev.Biol.—Animal, 50*, 851-857.

Portela, G. S., Azoubel, R., & Batigalia, F. (2007). Effects of Aspartame on Maternal-Fetal

and Placental Weights, Length of Umbilical Cord and Fetal Liver: A Kariometric

Experimental Study. *Int. J. Morphol, 25*(3), 549-554.

Position of the American Dietetic Association. (2004, January). Use of nutritive and

nonnutritive sweeteners. The American Dietetic Association Reports. *ournal of

American Ditetetic Association*, 255-275.

Public Health Service Food and Drug Administration. (1981). Department of Health and

Human Services. *Federal Reglster , 46*(142), 38284-38308.

Randolph, W. F. (1973). Notice offihing of petition for food additive. *Federal Registry, 38*,

5921.

Rangan, M. R., Kwan, J., Flood, V. M., Yu Louie, J. C., & Gill, T. P. (2011). Changes in 'extra'

food intake among Australian children between 1995 and 2007. *Obesity Research

and Clinical Practice, 5*, 55-63.

Ranney, R., & Oppermann, J. (1979). A review of the metabolism of the aspartyl moiety of

aspartame in experimental animals and man. *J. Environ. Pathol. Toxicol., 2*, 979-

985.

Rencüzoğulları, E., Ayaz Tüylü, B., Topaktaş M, İla, H. B., Kayraldız, A., Arslan, M., & Budak

Diler, S. (2004). Genotoxicity of Aspartame. *Drug and Chemical Toxicology, 27*(3),

257-268.

Renwick, A. G. (2006). The intake of intense sweeteners – an update. *Food Additives and

Contaminants, 23*(4), 327-338.

Sanyude, S., Locock, R. A., & Pagliaro, L. A. (1991). Stability of aspartame in water: organic

solvent mixtures with different dielectric constants. *J Pharm Sci, 80*(7), 674-676.

Sarett, H. P. (1973). Industry's Experience with Special Dietary Foods. *FOOD DRUG

COSMETIC LAW JOURNAL, 31*, 15730-15736.

Sasaki, Y. F., Kawaguchi, S., Kamaya, A., Ohshita, M., Kabasawa, K., Iwama, K., . . . Tsuda, S.

(2002). The comet assay with 8 mouse organs: results with 39 currently used food

additives. *Mutation Research, 519*, 103-119.

Sathyapalan, T., Thatcher, N. J., Hammersley, R., Rigby, A. S., Pechlivanis, A., Gooderham, N.

J., . . . Courts, F. (2015). Aspartame Sensitivity? A Double Blind Randomised

Crossover Study. *PLoS One, 10*(5), e0126039.

SCF. (1980). *Guidelines for the safety assessment of food additives (opinion expressed on 22 February 1980).Reports of the Scientific Committee on Food (10thseries). SCF; 1980. EUR 6892.Available from:http://ec.europa.eu/comm/food/fs/sc/scf/reports/scf_reports_10.pdf.* Luxembourg: Commission of the European Communities.

SCF. (1985). *Sweeteners (opinion expressed on 14 September 1984). Reports of the Scientific Committee on Food (16th series). SCF; 1985. EUR 10210 EN. Availablefrom: http://ec.europa.eu/comm/food/fs/sc/scf/reports/scf_.* Luxembourg: Commission of the European Communities.

SCF (Scientific Committee on Food), 1985. Sweeteners. Reports of the Scientific Committee for Food (Sixteenth Series), EUR 10210 EN, Commission of the European Communities, Luxembourg.

SCF. (1989). *Sweeteners (opinion expressed on 11 December 1987 and 10 November 1988). Reports of the Scientific Committee on Food (21st series). EUR 11617 EN. Available from: http://ec.europa.eu/comm/food/fs/sc/scf/reports/scf_reports_21.pdf.* Luxembourg:: SCF.

SCF (Scientific Committee on Food), 1989. Sweeteners. Reports of the Scientific Committee for Food (Twenty-first Series), EUR 11617 EN, Commission of the European Communities, Luxembourg.

SCF (Scientific Committee on Food), 1997. Minutes of the 107th Meeting of the Scientific Committee for Food, held on 12–13 June 1997 in Brussels. <http://europa.eu.int/comm/food/fs/sc/oldcomm7/out13_en.html>.

SCF (Scientific Committee on Food), 2002. Opinion of the Scientific Committee on Food: Update on the Safety of Aspartame (expressed on 4 December 2002).

<http://ec.europa.eu/food/fs/sc/scf/out155_en.pdf>.

SCF. (2001). *Guidance on submission for food additive evaluations Available from:* *http://ec.europa.eu/comm/food/fs/sc/scf/*. SCF.

SCF. (2002). *Update on the safety of aspartame.* 2002 tarihinde http://europa.eu/comm/food/fs/sc/scf/outcome_en. adresinden alındı

Schernhammer, E. S., Bertrand, K. A., Birmann, B. M., Sampson, L., Willett, W. C., & Feskanich, D. (2012). Consumption of artificial sweetener– and sugar-containing soda and risk of lymphoma and leukemia in men and women. *Am J Clin Nutr, 96,* 1419-1428.

Schulpis, K. H., Papassotiriou, I., Parthimos, T., Tsakiris, T., & Tsakiris, S. (2006). The effect of L-cysteine and glutathione on inhibition of Na-K-ATPase activity by aspartame metabolites in human erythrocyte membrane. *Eur J Clin Nutr, 60*(5), 593-597.

Scriver, C.R., Byck, S., Prevost, L., Hoang, L. (1996) The phenylalanine hydroxylase locus: a marker for the history of phenylketonuria and human genetic diversity. PAH Mutation Analysis Consortium, *Ciba Found Symp* 197, 73–90.

Scriver, C. R., & Kaufman, S. (2001). *Hyperphenylalaninemia: phenylalanine hydroxylase deficiency.* (C. R. Scriver, A. L. Beaudet, W. S. Sly, D. Valle, B. Childs, K. Kinzler, & B. Vogelstein, Dü) New York: McGrawHill.

Shaw, D. M., Camps, F. E., & Eccleston, E. G. (1967). 5-Hydroxytryptamine in the Hind-Brain of Depressive Suicides. *The British Journal of Psychiatry, 113*(505), 1407-1411.

Shaywitz, B. A., Sullivan, C. M., Anderson, G. M., Gillepsie, S. M., Sullivan, B., & Shaywitz, S. E. (1994). Aspartame, Behavior, and Cognitive Function in Children With Attention Deficit Disorder. *Pediatrics, 93*(1), 70-75.

Smyth, T. R. (1983). The FDA's Public Board of Inquiry and the Aspartame Decision. *Indiana Law Journal, 58*(4), 627-649.

Soffritti, M., Belpoggi, F., Degli, E. D., Lambertini, L., Tibaldi, E., & Rigano, A. (2006). First experimental demonstration of the multipotential carcinogenic effects of aspartame administered in the feed to Sprague-Dawley rats. *Environ Health Perspect, 114*, 379-385.

Soffritti, M., Belpoggi, F., Esposti, D. D., Lambertini, L., Tibaldi, E., & Rigano, A. (2006). First Experimental Demonstration of the Multipotential Carcinogenic Effects of Aspartame Administered in the Feed to Sprague-Dawley Rats. *Environmental Health Perspectives, 114*, 379-385.

Soffritti, M., Belpoggi, F., Tibaldi, E., Esposti, D. D., & Lauriola, M. (2007). Life-Span Exposure to Low Doses of Aspartame Beginning during Prenatal Life Increases Cancer Effects in Rats. *Environmental Health Perspectives, 115*(9), 1293-1297.

Spiers, P. A., Sabounjian, L., Reiner, L., Myers, D. K., Wurtman, J., & Schomer, D. L. (1998). Aspartame: neuropsychologic and neurophysiologic evaluation of acute and chronic effects. *Am J Clin Nutr, 68*, 531-537.

Spingola, D. (2015). *Screening Sandy Hook: Causes and Consequences.* Trafford publishing.

Stanley, M., & Stanley, B. (1990). Postmortem evidence for serotonin's role in suicide. *J Clin Psychiatry, 51*, 29-30.

Stegink, L. D. (1984). *Aspartate and glutamate metabolism.* (L. D. Stegink, & L. J. Filler, Dü) New York: 270 Madison Avenue.

Stegink, L. D. (1987). The aspartame story: a model for the clinical testing of a food additive. *American Journal of Clinic Nutrition, 46*, 204-215.

Stegink, L.D., Filer, L.J Jr., Bell, E.F., Ziegler, E.E., Tephly, T.R., et al. (1990) Repeated ingestion of aspartame-sweetened beverages: further observations in individuals heterozygous for phenylketonuria. *Metabolism* 39(10), 1076–1081.

Stokes, A. F., Belger, A., Banich, M. T., & Taylor, H. (1991). Effects of acute aspartame and acute alcohol ingestion upon the cognitive performance of pilots. *Aviat Space Environ Med, 62*(7), 648-653.

Suomi, S. J. (1984). *Effects ofaspartame on the learning test performance of young stumptail macaques.* (L. D. Stegink, & L. J. Filler, Dü) New York: Marcel Dekker.

TC Şeker Kurumu. (2015, Mart 25). *Yüksek Yoğunluklu Tatlandırıcıların Kurumumuzca İzlenmesi ve Denetlenmesi ile İlgili Basın Açıklaması.* http://www.sekerkurumu.gov.tr/haberler/yuksek-yogunluklu-tatlandiricilarin-kurumumuzca-izlenmesi-ve-denetlenmesi-ile-ilgili-basin-aciklamasi adresinden alındı

Tephly, T. R., & McMartin, K. E. (1984). Methanol metabolism and toxicity. *New York: Marcel Dekker*, 111-140.

Terphly, T. (1991). The toxicity of methanol. *J Life Sci, 48*(11), 1031-1041.

Tey, S., Salleh, N., Henry, J., & Forde, C. (2017). Effects of aspartame-, monk fruit-, stevia- and sucrose-sweetened beverages on postprandial glucose,insulin and energy intake. *International Journal of Obesity, 41*, 450-457.

TGK. (2002). *Türk Gıda Kodeksi (Tebliğ No:2002/ 56).* Ankara: Tarım ve Köyişleri Bakanlığı, Sağlık Bakanlığı.

Toledo, M. C., & Ioshi, S. H. (1995). Potential intake of intense sweeteners in Brazil. *Food Addit. Contam, 12*, 799-808.

Tüfekçi Alphan, M. E. (2014). Diabetes Mellitus ve Beslenme Tedavisi. M. E. Tüfekçi Alphan içinde, *Hastalıklarda Beslenme Tedavisi* (2. b., s. 415-507). Ankara: Hatiboğlu Basım ve Yayım San. Tic. Ltd. Şti.

U.S Food and Drug Administritation. (2015, May 26). *Additional Information about High-Intensity Sweeteners Permitted for use in Food in the United States.* 5 26, 2015 tarihinde U.S. Department of Health and Human Services: http://www.fda.gov/food/ingredientspackaginglabeling/foodadditivesingredients/ucm397725.htm adresinden alındı

Van den Eeden, S., Koepsell, T., Longstreth, W., van Belle, G., Daling, J., & et al. (1994). Aspartame ingestion and headaches: a randomized crossover trial. *Neurology, 44*, 1787–1793.

Vandevijvere , S., Chow, C. C., Hall, K. D., Umali, E., & Swinburn, B. A. (2015;93). Increased food energy supply as a major driver of the obesity epidemic: a global analysis. *Bull World Health Organ*, 446-456.

Virtanen, S. M., Rassanen, L., Paganus, A., Varo, P., & Akerblom, H. K. (1988). Intake of

    sugars and artificial sweeteners by adolescent diabetics. *Nutr. Rep. Int, 38*, 1211-

    1218.

Watts, R. S. (1991). Aspartame, headaches and beta blockers. *Headache, 31*, 181-182.

Whitehouse, C., Boulatta, J., & McCauley, L. (2008). The potential toxicity of artificial

    sweeteners. *Journal of the American Association of Occuoational Health Nurses, 56*,

    251-259.

WHO. (1987). *Environmental Health Criteria 70: Principles for the safety assessment of food

    additives and contaminants in food.* Geneva: WHO.

Wolraich, M. L., Lingdren, S. D., Stumbo, P. J., Stegink, L. D., Appelbaum, M. I., & Kırıtsy, M.

    C. (1994). Effects of Diets High in Sucrose or Aspartame on the Behavior and

    Cognitive Performance of Children. *The New England Journal of Medicine, 330*(5),

    301-306.

World Health Organization. (2008). *Global strategy on diet physical activity and health.*

Wurtman, R. J., & Maher, T. J. (1987). Effects of oral aspartame on plasma phenylalanine in

    humans and experimental rodents. *Journal of Neural Transmission, 70*, 169-173.

Yarmolinsky, J., Duncan , B. B., Chambless, L. E., Bensenor, I. M., Barreto, S. M., Goulart, A.

    C., . . . Schmidth, M. I. (2015). Artificially Sweetened Beverage Consumption Is

    Positively Associated with Newly Diagnosed Diabetes in Normal-Weight but Not in

    Overweight or Obese Brazilian Adults. *The Journal of Nutrition*, doi: 10.3945/

    jn.115.220194.

Zehetner, A., & McLean, M. (1999). Aspartame and the internet. *Lancet, 354*, 78.

YOUR KNOWLEDGE HAS VALUE

- We will publish your bachelor's and
  master's thesis, essays and papers

- Your own eBook and book -
  sold worldwide in all relevant shops

- Earn money with each sale

Upload your text at www.GRIN.com
and publish for free